Going Global?
A study of our interconnected world

Catherine Owen

GCSE Geography Teachers' Toolkit
Series editors: Ruth Totterdell and Justin Woolliscroft

Geographical Association

Editors' introduction

All GCSE specifications being taught from 2009 have been written to reflect QCDA's subject criteria for geography. These state that 'GCSE specifications in geography should encourage learners to be inspired, moved and changed by following a broad, coherent, satisfying and worthwhile course of study' and that students should 'actively engage in the process of geography to develop as effective and independent learners, and as critical and reflective thinkers with enquiring minds'.

The Geographical Association's Manifesto *A different view* (*www.geography.org.uk/adifferentview*) consolidates this view: 'Living Geography brings contemporary context and real world enquiry to the curriculum'. One thing is certain: school geography is undergoing significant change at all key stages, which affords us geographers many opportunities to revitalise and reinvigorate the geography that is taught in schools.

The *GCSE Geography Teachers' Toolkit* is designed to help teachers develop meaningful schemes of work from their chosen GCSE specification. It takes a refreshing look at how GCSE geography can be taught so that students become enthused, engaged and motivated by it – thinking and acting as geographers.

The series covers six themes, each featuring in all the GCSE specifications:

- sustainable development
- uneven development
- globalisation
- climate change
- waste
- extreme environments.

Each title in the series provides a unit of work: a bank of lesson ideas and accompanying downloadable resources.

Each **lesson** includes key questions to engage learners, and provides opportunities for them to practise the skills (including a decision making exercise) required to tackle GCSE examinations. It also offers guidance on how you can incorporate Living Geography approaches in your GCSE teaching. The lessons are designed to be thought-provoking and we hope you will be encouraged to use the suggested teaching strategies elsewhere in the curriculum.

The supporting **resources** – maps, PowerPoints, photo sets, information sheets, activity sheets and weblinks – can be used directly in the classroom or adapted to suit your needs and those of your students. The resources can be downloaded from a password-protected area on the Geographical Association website. Details of how to access these resources are given inside the front cover.

The *GCSE Geography Teachers' Toolkit* series is the work of an experienced and respected team of authors, each responding creatively and imaginatively, and in a range of teaching styles, to the potential of the revised GCSE specifications. As with the hugely successful *KS3 Geography Teachers' Toolkit* series, it is a rich source of teacher-to-teacher advice – a conversation between professionals about how best to make the most of these new opportunities.

Ruth Totterdell and Justin Woolliscroft, 2010

Contents

Teaching about globalisation

Globalisation is not a new process; it has been with us since the rise of capitalism. However, the rapid pace and all-consuming nature of globalisation are particular relevant in the twenty-first century. Moreover, although its effects are unevenly distributed across the globe, almost everyone is affected by globalisation in some way.

The resulting global interdependence of countries, economies and cultures means that actions taken in one place often have repercussions elsewhere in the world. Therefore, what is important to us as teachers is that our students understand their place in this globalised world. To achieve that, we need to understand what the term means.

> *'Globalisation can be seen as a positive, negative or even marginal process. And regardless of whether it works for good or ill, globalisation's exact meaning will continue to be the subject of debate among those who oppose, support or simply observe it.*
>
> *'A recent report in the Press Gazette ... dealt with attempts by a BBC focus group to throw some light on how far television audiences understand news reports.*
>
> *'In one clip, economics editor Evan Davis referred to "Globalisation – whatever that means". A panellist replied: "Well if he doesn't know what it means, how the hell are we supposed to?"'* [1]

It is somewhat daunting: if a respected economics journalist is unsure what globalisation means, how are we to go about teaching it? This unit offers a range of strategies and contexts to engage students meaningfully with the subject matter. From examining different definitions of the term globalisation students can quickly move on to look at cultural, economic, environmental and political globalisation.

A shop in Kibera, Kenya, selling Coca-Cola. Photo: Kevin Cook

The economic effects of globalisation are often the focus of media attention (e.g. the BBC and *Guardian* weblinks – see p. 35 for details), whereas its cultural effects can be of more interest to students. For instance, the image of Kibera (a shanty town near Nairobi, Kenya) shown above indicates that you can buy Coca-Cola almost anywhere in the world. This can generate a number of questions: Is this type of globalisation good or bad? Does the sale of Coca-Cola just mean a wide range of drinks locally, or is it corrupting local cultures? Might it actually strengthen local cultures? (see Yale Global weblink).

Conveying the relationship between cultural and economic geography can be a challenge. But we can start with students' personal experiences and build out from there. A straightforward question like 'Where were your clothes made?' followed by a survey of labels enables us to link the students with other countries. A discussion of the global division of labour (with different countries involved at different stages – see e.g. CDRB weblink) will rapidly widen the scope and depth of students' understanding of issues at local, national and global levels.

At the heart of discussions on economic globalisation lie questions of who holds power. Is it in the hands of trans-national corporations (TNCs) or national governments? The World Trade Organisation (WTO) plays a major role in regulating trade between states, but is often accused of bias towards the interests of TNCs (see WTO and Global issues weblinks). Here simulation and role play can motivate the students to consider issues of the balance of trade between countries and trade blocs.

When students ask 'Can individuals make a difference?', we can point to anti-globalisation movements (motivated by, for example, environmentalism, child labour or anti-capitalism), which have developed to counter what they perceive as globalisation's negative effects. Students can unpick different people's views (pro, anti or otherwise) and become aware that the different lenses people use to view globalisation are shaped by that individual's contexts and experiences. Thus, the students' own knowledge and understanding of globalisation will shape their views of the probable and their preferred futures.

1 From: 'What is Globalisation', Simon Jeffery, *Guardian*, 31 October 2002. Available online at *www.guardian.co.uk/world/2002/oct/31/globalisation.simonjeffrey*.

Living Geography and GCSE

Geography in schools is concerned with perceptive, deep description of the real world. It seeks explanations about how the world works and helps us to think about alternative futures. This is *Living Geography* – a geography that:

- embraces young people's geography and experiences
- is current and future oriented
- has a scale zoom lens, so that the local is always set in a global context
- raises questions of change, sustainability and development.

The strength of 'thinking geographically' is that it brings school geography alive – young people ask questions about and investigate their own world. To varying extents, all of the GCSE specifications reflect *Living Geography*, as the following extracts indicate.

WJEC

- Relevance of study to students' everyday lives.
- An emphasis on 'geographical futures' which will encourage students to apply their understanding of current trends to identify alternative future scenarios and evaluate their sustainability.

AQA

- To produce students who are aware of their responsibilities to other people, the environment and the sustainability of the planet.
- To build a solid geographical foundation through popular and familiar topics, with inspiring and motivating content to raise students' achievement.

CEA

- To help students develop as individuals and as contributors to society, as well as to the economy and environment.
- To enable students to actively engage in the process of geography to develop as effective and independent learners.

Edexcel

- An increased necessity to teach a specification which reflects the world we live in.
- Importance of geography in the twenty-first century: climate change; sustainability; our world in the media spotlight.

OCR

- To appreciate the importance of geography for understanding the world, supporting the learning of geographical concepts and encouraging candidates to think geographically.
- To engage students actively in the process of geography, so that they think like geographers and develop their responsibilities as global citizens.

The *GCSE Geography Teachers' Toolkit* series covers six themes – sustainable development, uneven development, globalisation, climate change, waste and extreme environments – which feature in all specifications. Rather than looking at every facet of each of these big ideas, the series exemplifies how they can be taught through *Living Geography* approaches. Teachers can use the materials to develop a GCSE teaching programme that is rigorous, engaging and relevant for students in the twenty-first century.

Going Global? A study of our interconnected world

When they first encounter globalisation, students may see it as dry and boring; not a topic that has direct relevance to their lives. However, as they explore the issues involved they begin to see that not only are their lives directly affected by globalisation, but also that they can play a part in shaping the impact that globalisation will have in the future.

Lesson coverage

The first four lessons focus on how students' lives are affected by globalisation. They build on their prior understanding of global issues and the relevance of globalisation to their lives. In these lessons students have the opportunity to learn about and reflect on:

- their place in a globalised world
- the global division of labour involved in manufacturing the goods they buy
- the role of the internet in their lives and in globalisation
- how TNCs make a big difference to their lives and the lives of others around the world.

Students are encouraged to discuss definitions of and views about globalisation, considering why different people hold very different (and sometimes very strong) opinions about this issue.

The next four lessons move from the personal to consider experiences of cultural and economic globalisation in countries at different stages of development. A lesson looking at inequalities associated with world trade is followed by a lesson each on China, the UK and Kenya, in which students consider:

- how globalisation has led to change in China
- how UK trade benefits from EU protection
- how the impacts of globalisation on Kenya could be managed to maximise the benefits and minimise the problems.

Thus, the lens is widened from the local to the national/global scales, while allowing students to think about change, sustainability and development.

Lessons 9 and 10 comprise a synoptic decision-making exercise widening the field of study further - to the global scale. This gives students the opportunity to consider what the world is likely to be like in 2050, what they would prefer it to be like in 2050, and what they need to do to make their preferred future a reality.

Tackled through this *Living Geography* approach, the topic of globalisation will inspire and motivate students to develop as effective and independent learners.

GCSE examination skills

All GCSE specifications suggest a range of skills that should be practised throughout the duration of the course.

According to the GCSE specifications, students should be able to:

- identify relevant geographical questions
- plan a sequence of enquiry
- carry out fieldwork and out-of-classroom learning
- collect appropriate primary and secondary data
- draw diagrams, maps, graphs and sketches

- describe, analyse and interpret evidence from written articles, tables, charts, graphs, diagrams and cartoons
- describe, analyse and interpret evidence from maps at various scales and of varying types
- describe, analyse and interpret evidence from photographs (ground, aerial and satellite imagery)

- make informed geographical decisions
- draw and justify conclusions
- evaluate their work and discuss the validity and limitations of their evidence and conclusions
- formulate and justify an argument
- communicate information in ways suitable for a range of target audiences
- use new technologies, including GIS
- acquire and use geographical vocabulary
- use ICT to research or present findings.

Throughout their GCSE course, students should be given opportunities to learn new skills and practise those they have already acquired. They should also be afforded opportunities to work independently, so that they can display and develop their initiative and creativity.

Each lesson in this book is related to how it will develop the students' exam skills. There are also some exam practice questions for each lesson. The questions and answers can be downloaded from *www.geography.org.uk/gcsetoolkit*.

The activities in this book are designed to give students the chance to develop their exam skills in a motivational and inspiring context. From the first lesson, they are encouraged to see globalisation as a controversial issue and to consider how and why different stakeholders view globalisation differently. Students explore how their lives and the lives of people around the world are connected and interdependent.

A key aim of these lessons is to develop students as independent learners. There are opportunities for enquiries both within and outside lessons, and students are required to make informed geographical decisions in different contexts. Different ways of learning are explored, including de Bono's coloured hats and concept mapping; students will find out which learning styles best suit them and make use of these techniques in other contexts, such as when preparing for controlled assessments or revising for exams.

Medium term plan

Lesson/key questions	Learning objectives	Teaching and learning activities	Resources	Exam preparation opportunities
Lesson 1: What is my place in the globalised world? How is my life linked to other countries? What impact do these links have on other countries? What does globalisation mean to me and to other people?	Know that our lives are linked to those of others living around the world Consider the impact that these links have on other countries Be able to annotate a world map to show links Understand that globalisation means different things to different people	Students think of ways in which they are linked to other countries and which countries they have links with Students annotate a world map to show these links and start to consider the impact these links may have on the countries identified Students discuss what globalisation means to them and compare their definitions with others, thinking about why different people view globalisation in different ways	PowerPoint 1 Map 1 Photo set 1	Students look at globalisation in the context of their own lives, which emphasises the importance of their personal geography and experiences Widening the scale sets the local in a global context
Lesson 2: Who made my iPod? Where do the components of an iPod come from? Is there a global division of labour in its production?	Know that items we purchase are made in different stages Understand that different stages in the industrial process tend to take place in different places in the world Consider the impact that the global division of labour may have on people and economies in different parts of the world	Students explore the stages involved in making an iPod using a card sort activity Students are introduced to the concept of a global division of labour Students research where different stages in the global division of labour are likely to take place, identifying any patterns that emerge Students consider the implication of these patterns on people in different parts of the world	Information sheet 1 Activity sheet 1 PowerPoint 2 Weblinks 1 and 2	Students use atlases and web-based maps to describe, analyse and interpret evidence from maps at various scales and of varying types
Lesson 3: Has the internet brought us closer together? How does the internet connect us to the rest of the world? Does the internet have a positive impact on globalisation?	Know that the internet has been a major force in globalisation Understand that the internet can be used in both positive and negative ways Reflect upon the role of the media and the internet in cultural globalisation	Students reflect upon how they use the internet and how it connects them to the rest of the world Students work in pairs to study images and text related to the internet and globalisation. They produce a statement answering the key question based upon what they have read Complete a 2-4-8 activity, then share ideas as a class and discuss issues raised. Students reflect on difficulties in producing their statements and how they were resolved	Information sheets 2 and 3 Activity sheet 2 PowerPoint 3 Weblinks 3-6	Students develop their skills of decision making, presenting an argument and negotiation through group work This work also contributes to their 'team workers' requirement in the personal learning and thinking skills

Lesson/key questions	Learning objectives	Teaching and learning activities	Resources	Exam preparation opportunities
		Students reflect on the role of the media and the internet in cultural globalisation		
Lesson 4: What do TNCs mean to me? What is a TNC? How do TNCs impact on people's lives? Why do people perceive this impact differently?	Know what TNCs are Be able to use de Bono's coloured hats to investigate a TNC Understand that different people view TNCs in different ways	True/false activity to introduce TNCs Students work in groups to carry out an internet investigation into one TNC. Each group takes on the perspective of one of de Bono's six coloured hats Groups share findings and complete notes grid, then produce individual responses to the key question Students evaluate de Bono's coloured hats as a framework for learning	Information sheet 4 Activity sheet 3 PowerPoint 4 Photo set 2 Weblinks 7–9	Students plan a sequence of enquiry and collect secondary data using ICT
Lesson 5: Balanced trade or trade balance: is the balance right? What is meant by the term 'balance of trade'? Why do some countries have a positive trade balance and others a negative one?	Know what is meant by 'balance of trade' Understand that a country's role in the global division of labour will influence its balance of trade	Students look at trade figures for different countries Students carry out a simulation activity on the impact of world trade on the balance of trade of countries at different stages of development Students reflect upon what they have learnt from the simulation and how it could be developed	Information sheets 5-7 Activity sheet 4 PowerPoint 5 Weblinks 10 and 11	Students develop their ability to make informed geographical decisions. This also contributes to their development as 'effective participators' in relation to personal learning and thinking skills
Lesson 6: How is globalisation changing China? What was China like in the past? What are the effects of globalisation on China and its people?	Know where China is and some back-ground information about it Understand that globalisation has led to significant change in China Be able to present information using a concept map	Students explore issues of globalisation in China via images Students produce a concept map covering political, economic, social and environmental aspects of globalisation China	Information sheet 8 Activity sheet 5 PowerPoint 6 Photo set 3 Weblinks 12–14	Students' use of concept maps allows them to explore how different issues are linked and to think creatively. They can then apply this technique to revise for a topic or issue
Lesson 7: Does the UK need the protection of the EU? What is the EU and who are its members? What is a trading bloc and how does it work?	Know what the EU is and which countries are members Understand how the EU operates as a trading bloc to protect the trade of member countries Understand that actions taken by	Students list members of the EU Students are introduced to the EU sugar trade and consider how this trade should be organised to best benefit the UK in relation to use of subsidies, tariffs and taxes Students reflect on how these decisions would affect people and places outside of the EU. They draw conclusions about whether protectionism is a good or a bad	Information sheet 9 Activity sheet 6 PowerPoint 7 Weblinks 15-17	Students develop areas of geographical vocabulary that they may not encounter every day Students must also make informed geographical decisions on some aspects of globalisation

Lesson/key questions	Learning objectives	Teaching and learning activities	Resources	Exam preparation opportunities
How does being an EU member protect the UK's trade? How does this affect trade in other parts of the world?	the EU affect people and places elsewhere	thing for the UK, EU and the world		
Lesson 8: Are Kenyans winners or losers in the globalisation game? What is life like for people in Kenya? How does globalisation impact upon their daily lives? How can globalisation be managed to benefit places like Kenya?	Know where Kenya is and some background information about it Understand that different people in Kenya experience globalisation in different ways Understand that the benefits of globalisation can be maximised and the problems minimised with careful management	Pairs of students are allocated role cards showing information about the lives of different Kenyans and the impact globalisation has on their lives Students arrange themselves into a continuum to show who has the most positive experience of globalisation and who the least Students discuss the images and questions on PowerPoint 8 Students answer an exam style question	Information sheets 10-13 Activity sheet 7 PowerPoint 8 Photo set 4 Weblinks 18–21	Students apply the knowledge, understanding and skills they have developed over previous lessons to an exam-style context
Lesson 9: What will our global future be? Part 1 What did people imagine our future would be like? What is the most likely global future? What future would you prefer?	To understand the different ways that the future can be envisioned To understand that decisions made today will influence the way globalisation affects us in the future	Students read excerpts from utopian and dystopian texts about what the future may hold for the world. They discuss their own views about what the future holds Students work in small groups to produce their view of the most likely global future, their preferred global future and the steps needed to make the preferred global future come true. This is in preparation for a presentation in the final lesson	Activity sheet 8 PowerPoint 9 Photo set 5 Weblinks 22-25	Students have the opportunity to show their geographical knowledge and understanding in preparing their presentations. They also need to draw and justify conclusions In choosing the format of their presentation, students have the opportunity to try out different approaches to their work
Lesson 10: What will our global future be? Part 2 How do our global futures compare? What are the strengths of our preferred global futures? What are the weaknesses of our preferred global futures?	To understand that decisions made today will influence the way globalisation affects us in the future	Small groups of students present their preferred global future and the steps needed to make it come true Students peer-assess other groups' presentations using stars and wishes criteria, and self-assess their own presentation	Activity sheet 9 PowerPoint 10 Weblink 26	During the presentations, students will need to draw and justify conclusions and present their arguments to an audience and through controlled assessment Peer- and self-assessment offers students the opportunity to discuss the validity and limitations of their conclusions. It also contributes to 'reflective learners' in personal learning and thinking skills

Lesson 1

What is my place in the globalised

Key questions

- How is my life linked to other countries?
- What impact do these links have on other countries?
- What does globalisation mean to me and to other people?

Learning objectives

- To understand that our lives are linked to those of people who live in other countries
- To consider the impact that these links have on people in other countries
- To be able to annotate a world map to show my links to other countries
- To understand that globalisation means different things to different people

Key terms (see glossary)

- cultural (globalisation)
- economic (globalisation)
- globalisation

Online resources

- **PowerPoint**
  1 What is my place in the globalised world?
- **Map**
  1 Outline world map
- **Photo set**
  1 Clothing labels

Stimulus

Display the second slide in **PowerPoint 1** as students enter the room. It shows Jake, a global teenager, and is labelled to indicate how he is linked to different countries around the world. Invite students to draw a quick self-portrait with labels to show how they are linked to different countries. Share some of the portraits.

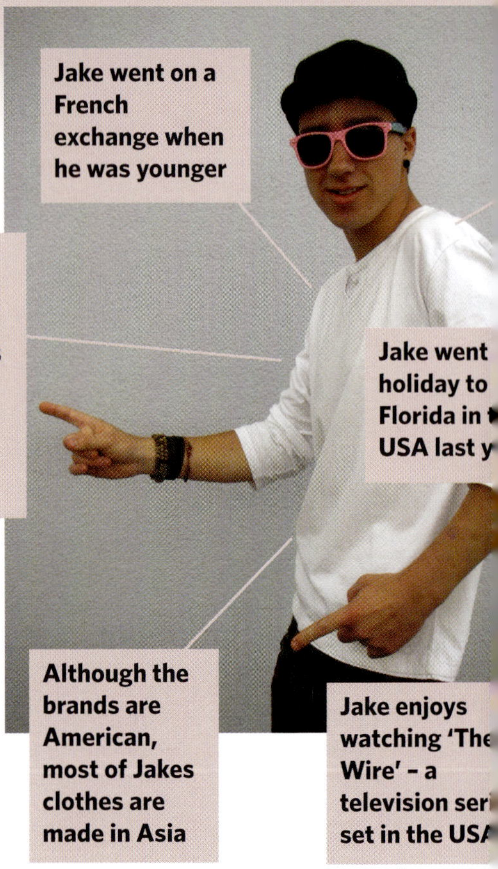

Development

Students use atlases to identify the countries they have links with, then annotate a copy of **Map 1** to show how they are linked to these countries (as modelled on **PowerPoint 1**). Ask students to suggest the impacts that their links may have, e.g. 'My school bag is made in China. This means that Chinese workers have jobs.'

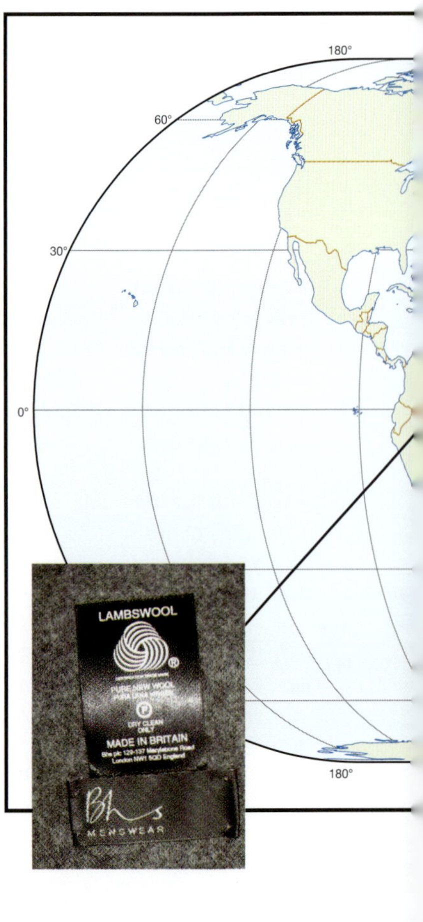

Consolidation

Introduce the students to the term 'globalisation'. One dictionary defines it as 'when available goods and services, or social and cultural influences, gradually become similar in all parts of the world'. Discuss the following questions:

- What are the differences between cultural and economic globalisation?

d?

Jake's grandfather is Maori – from New Zealand

Photo: Catherine Owen

- What does globalisation mean to you?
- What impact does globalisation have on your life?
- Do you think that globalisation is a good or a bad thing?

Compare the different definitions/quotes about globalisation on **PowerPoint 1** with the students' definitions. Why do they think different people view globalisation in different ways?

Extension

Either

Students look at labels on five or ten of their favourite items of clothing (see **Photo set 1**) to find out whether they have a global wardrobe. What impact has the fashion industry had on them? What about other parts of the world?

Or

they think about family holidays. Where have they visited in the last five years? They could ask their parents and grandparents about their childhoods: where did they go on holiday? How does this illustrate the recent rapid expansion of globalisation?

Living Geography

Looking at globalisation in the context of the students' own lives emphasises the importance of their personal geography and experiences. Widening the scale from their self-portrait to their global links sets the local in a global context.

Exam skills

Annotating a world map develops students' awareness of where different countries are located in the world, as well as developing their map skills. The classroom discussions will enable students to develop their geographical vocabulary.

Exam practice

a What does globalisation mean? Give one example to support your answer. (2 marks)

b Explain why globalisation can be viewed in both positive and negative ways. Give examples to support your answer. (4 marks)

(Answers on **Exam practice questions and answers** download.)

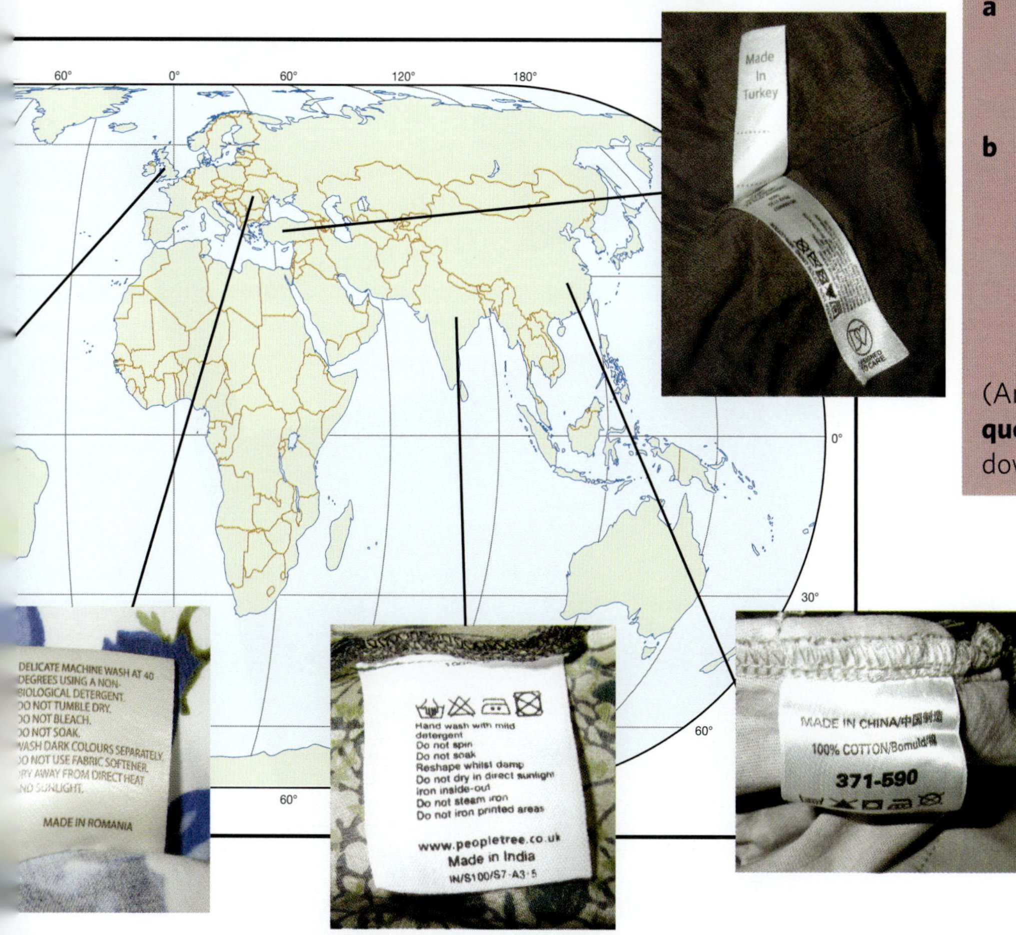

Lesson 2

Who made my iPod?

Key questions

- Where do the components of an iPod come from?
- Is there a global division of labour in its production?

Learning objectives

- To know that items we purchase are made in different stages
- To understand that different stages in the industrial process tend to happen in different places in the world
- To consider the possible impact of the global division of labour on people and economies in different parts of the world

Key terms (see glossary)

- global division of labour
- less economically developed countries (LEDCs)
- more economically developed countries (MEDCs)
- newly industrialised countries (NICs)
- primary sector
- quaternary sector
- secondary sector
- tertiary sector

Online resources

- **Information sheet**
  1 Using Worldmapper
- **Activity sheet**
  1 Where does my iPod come from?
- **PowerPoint**
  2 Who made my iPod?
- **Weblinks**
  1 Worldmapper
  2 Global gang

Stimulus

Using **PowerPoint 2** explain that when the 30GB Video iPod was launched in 2007 it sold for US$299. Then pose the questions: Where was it made? And how much of the retail price was 'spent' at each stage of its production? Students carry out a sequencing activity to explore the stages involved in making an iPod using the cards on **Activity Sheet 1**.

Students can decide whether each stage involves primary, secondary, tertiary or quaternary industries. They could also decide whether the 'cost' of each stage of production is fair and/or why some companies have their headquarters in one country and manufacture their products in other countries.

Development

Introduce the concept of a global division of labour. This is a way of organising work (especially making things) as a set of separate processes, each carried out by different (groups of) people. Does the iPod case study indicate that there is a global division of labour? How does this compare with the way products were made 50 years ago?

Show students atlases and/or **Weblink 1** and ask how they could use these resources to find out where the different stages of the industrial process take place. **Information sheet 1** explains how they should use **Weblink 1** to research which countries are mainly involved in which industries. Students consider whether these countries are more or less economically developed or newly industrialised countries (MEDCs, LEDCs or NICs) and identify patterns (see examples on **PowerPoint 2**). Some students could also consider how the global recession has affected these patterns.

Consolidation

Ask students to consider the following questions:

- How can the global division of labour be described?
- Do you think these patterns will stay the same or will they change in the future? Why?
- What impact do you think the global recession has had on these trading patterns?
- Do you think the global division of labour is a good thing or a bad thing? Why?

Extension

Students use **Weblink 2** - an animated game of snakes and ladders including questions about world trade - to find out more about the global division of labour and associated issues.

This Worldmapper map (below) shows electronic exports. Ireland, Taiwan, Malaysia, Hong Kong and Japan export the most. © Copyright 2006 SASI Group (University of Sheffield) and Mark Newman (University of Michigan).

Living Geography

Many students own an iPod, or know someone who does, so exploring the stages of its production taps into their own experiences and stimulates their interest. Again, moving between the iPod in their pocket and the global scale of production sets the local in a global context.

Exam skills

By using atlases and/or **Weblink 1** students will describe, analyse and interpret evidence from maps at various scales and of varying types.

Exam practice

With reference to an example that you have studied, describe the global division of labour. (4 marks)

(Answers on **Exam practice questions and answers** download.)

Lesson 3
Has the internet brought us closer together?

Key questions
- How does the internet connect us to the rest of the world?
- Does the internet have a positive impact on globalisation?

Learning objectives
- To know that the internet has been a major force in globalisation
- To understand that the internet can be used in both positive and negative ways
- To reflect upon the role of the media and the internet in cultural globalisation

Key terms (see glossary)
- interdependence
- internet
- media

Online resources
- **Information sheets**
  2 internet use extracts: group A
  3 internet use extracts: group B
- **Activity sheet**
  2 internet use and globalisation
- **PowerPoint**
  3 Has the internet brought us closer together?
- **Weblinks**
  3 Queens University
  4 Eric digests
  5 Geography of the internet
  6 Shift happens

Stimulus
Display the cartoon on **PowerPoint 3** as students enter the room. Students reflect in pairs on how they use the internet and how it connects them with the rest of the world. Widen into class discussion: What is positive about their use of the internet? What is negative?

Development
In pairs, students look at either **Information sheet 2** (which suggests that the internet is bringing people closer together) or **Information sheet 3** (which suggests the opposite). They come up with a statement answering the question: 'Has the internet brought us closer together?' and record this on **Activity sheet 2**. They must justify their argument by referring to their information sheet. Students then join into a four with a pair who have looked at the alternative sheet. Each pair presents their statement then the group of four negotiates a new statement in answer to the question. Two groups of four join together; again they present their statements and negotiate a new statement in answer to the question. Finally, the groups of eight read their statements to the class. This is then followed by a class discussion of issues that arise.

Consolidation
Students reflect on any difficulties they encountered in producing statements and how these were resolved. Referring back to **Lesson 1**, they could also reflect on the

Photo: Shaun Flannery

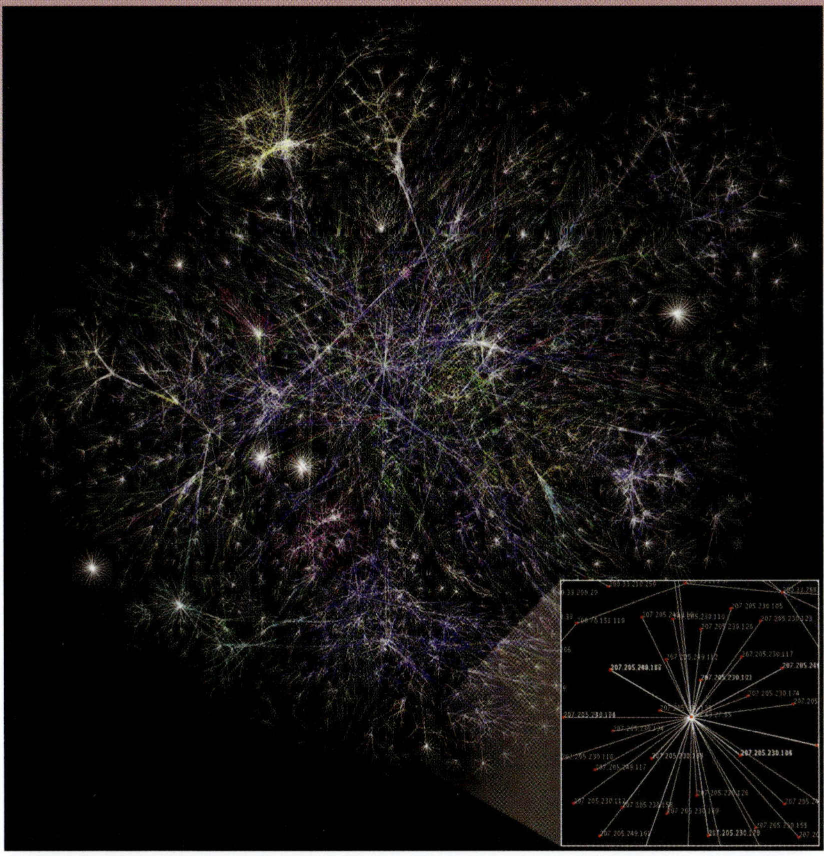

Partial map of the internet based on Opte Project data[1]*.*
Photo: ⓘ Matt Britt

role of the media and the internet in cultural globalisation.

Extension

Students watch the 'Geography of the internet' movie on **Weblink 5** (this includes a link to download ideas for lessons based around the movie). They could also watch the 'Shift happens' presentation and join in the online discussion of these issues on **Weblink 6**.

Note

This lesson includes discussion of sensitive and controversial issues, such as religious extremism. It can also consider why the authorities in some countries find the total freedom of internet use problematical (e.g. Myanmar/China). It is important that you establish ground rules for discussing such issues. **PowerPoint 3** includes some pointers. For more advice on controversial discussions in the classroom see **Weblinks 3 and 4**.

1 Partial map of the internet based on Opte Project (*www.opte.org*) data from 15 January 2005. Each line is drawn between two nodes, representing two IP addresses. The line lengths are indicative of the delay between the two nodes. This graph represents less than 30% of the Class C networks reachable by the data collection programme. Lines are colour-coded according to their corresponding RFC 1918 allocation as follows:

- dark blue: net, ca, us
- green: com, org
- red: mil, gov, edu
- yellow: jp, cn, tw, au, de
- magenta: uk, it, pl, fr
- gold: br, kr, nl
- white: unknown.

Living Geography

Again, students start with their own experiences of the internet then widen their lens to consider issues at a global scale. While globalisation is not a new phenomenon (some commentators say it dates back to Spanish conquistadors and their attempts to find El Dorado, if not further) – no one can deny that the internet is a major force in the current phase of globalisation.

Exam skills

Using the 2-4-8 technique (working in pairs, then groups of four and finally groups of eight) enables students to develop their skills of decision making, presenting an argument and negotiation. This also contributes to their 'team workers' requirement in the personal learning and thinking skills.

Exam practice

To what extent has new technology been important in accelerating the rate of globalisation? (5 marks)

(Answers on **Exam practice questions and answers** download.)

Lesson 4

What do TNCs mean to me?

Key questions

- What is a transnational corporation (TNC)?
- How do TNCs impact on people's lives?
- Why do people perceive this impact differently?

Learning objectives

- To know what transnational corporations (TNCs) are
- To be able to use de Bono's coloured hats to investigate a TNC (or issue) from different perspectives
- To understand that different people view TNCs in different ways

Key terms (see glossary)

- multinationals
- transnational corporations (TNCs)

Online resources

- **Information sheet**
  4 Labels for de Bono's hats
- **Activity sheet**
  3 Investigating TNCs using de Bono's hats
- **PowerPoint**
  4 What do TNCs mean to me?
- **Photo set**
  2 Kibera, Nairobi, Kenya
- **Weblinks**
  7 Instructables
  8 Guardian 'Academic experts'
  9 BBC special report

Stimulus

Students complete the true/false activity shown on the first slide **PowerPoint 4** as soon as they enter the room. Reveal answers – are there any surprises?

Development

Describe how transnational corporations (TNCs) operate, i.e. that they are business organisations that operate across national boundaries. Show the image of a shop selling Coca-Cola in Kibera in Kenya on **PowerPoint 4** (see also **Photo set 2**) to illustrate your explanation.

Explain that the whole class will investigate a TNC. Ask them to suggest which TNC they are most interested in and why they are interested in that particular one. Decide which TNC to investigate. Divide the class into six groups and give each group one of de Bono's coloured hats (**Information sheet 4** and **Weblink 7**). Use **PowerPoint 4** to explain the meaning of each hat and how this will affect each group's research. They should note what their hat means on **Activity sheet 3**.

Give students a set amount of time to carry out internet research on the TNC and complete the relevant section of their group's copy of **Activity sheet 3**. The groups then share their findings as a class, again noting key points on **Activity sheet 3**. Finally, they spend a few minutes working individually to produce their own response to the key question 'What do TNCs mean to me?'

Photo: Catherine Owen

A barber shop in Kibera, Kenya, pays homage to a well-known British football team. Photo: Kevin Cook

Consolidation

How useful are de Bono's six coloured hats as a framework for learning? Can students think of other issues that could be investigated in this way? **Weblinks 8 and 9** provide more information on de Bono's ideas.

Extension

Students note down how many TNCs they come across in one day. They could look at packaging on food and drink, the names of shops, advertising hoardings, etc. Has the growing influence of TNCs on world trade improved their quality of life? Would people from other countries feel the same way?

Note

If your lessons are short, this may be too much to fit into one session, so you may wish to divide the activities into a research lesson and a presentation lesson.

Living Geography

Initially, students probably will feel that TNCs are not part of their personal geography. However, once they find out more about the scope of these organisations, they will realise that TNCs have a huge impact on their lives.

Exam skills

In this lesson students plan a sequence of enquiry and collect secondary data using ICT.

Exam practice

With reference to an example, discuss the advantages and disadvantages that transnational corporations bring to LEDCs. (6 marks)

(Answers on **Exam practice questions and answers** download.)

Coca-Cola signs in Mae Sai, on the border between Thailand and Burma. Photo: Bryan Ledgard

Key questions

- What is meant by the term 'balance of trade'?
- Why do some countries have a positive trade balance and others have a negative one?

Learning objectives

- To know what is meant by 'balance of trade'
- To understand that a country's role in the global division of labour will influence its balance of trade

Key terms (see glossary)

- balance of trade
- exports
- global division of labour
- imports

Online resources

- **Information sheets**
  5 Sunbrella simulation: role cards
  6 Sunbrella simulation: country labels
  7 Sunbrella simulation: price list
- **Activity sheet**
  4 Sunbrella simulation: accounts sheet
- **PowerPoint**
  5 Global trade – is the balance right?
- **Weblinks**
  10 International Monetary Fund
  11 Oxfam 'Milking it'

Lesson 5

Balanced trade or trade balance: is t

Stimulus

As the students enter the classroom, display the first slide of **PowerPoint 5**, which shows a table of trade figures for different countries. Ask them what conclusions they can draw from the figures.

Development

Students take part in the simulation described on **Information sheet 5** and on **PowerPoint 5**. This simulation aims to develop their understanding of why different countries have different balances of trade and how global trade is influenced by such factors as trading blocs and natural disasters.

In groups representing different countries (**Information sheet 5**) the students are to produce the latest breakthrough in nano-technology – the **Sunbrella**! This amazing new product has a clear top with built-in ultra-violet filters to protect you from the sun's rays on sunny days, whereas on rainy days it turns opaque, revealing a fashionable pattern and making the wearer feel cosily sheltered from the rain. Each group's aim is to own as many Sunbrellas as possible, but they have to manufacture and transport them as follows:

- two groups make Sunbrella handles
- two groups make Sunbrella tops
- two groups apply the nano-technology and assemble the Sunbrellas
- two groups transport the components and finished Sunbrellas between countries

The brilliant Sunbrella!

balance right?

Photo: Catherine Owen

- one group provides insurance and charges for the use of the nano-technology for Sunbrellas.

Groups earn paper clips for their work and use the paper clips to purchase raw materials and finished Sunbrellas. They should keep a tally of their paper clip earnings using **Activity sheet 4**. (**Information sheet 7** indicates the price of activities.) Display the News Flash! slides from **PowerPoint 5** at intervals to add impetus to the simulation.

Consolidation

Students reflect upon what they have learnt from the simulation and how it could be developed to further enhance learning, including:

- Who 'won'? Why?
- Which countries had positive trade balances?
- Which countries had negative trade balances?
- What affects the trade balance of a country?
- The European Union is a trade bloc - how did this benefit Bulgaria?
- What else did you learn from this simulation?
- How could this simulation be improved/extended?

Individual students then produce an answer to the question: 'World trade - is the balance right?'

Extension

Students could find out more about world trade using the online simulations at **Weblinks 10 and 11**.

Living Geography

Having developed a global perspective in relation to local experiences in previous lessons, this lesson opens at the global scale and encourages students to think about the role of different countries and groups of countries within the global context. It raises questions of change, sustainability and development.

Exam skills

Students will develop their ability to make informed geographical decisions. This also contributes to their development as 'effective participators' in relation to personal learning and thinking skills.

Exam practice

What problems may world trading patterns cause for LEDCs? (4 marks)

(Answers on **Exam practice questions and answers** download.)

Lesson 6

How is globalisation changing China

Bailian Xijiao shopping centre, Shanghai. Photo: Sarah Maude

Key questions

- What was China like in the past?
- What are the effects of globalisation on China and its people?

Learning objectives

- To know where China is and some background information about it
- To understand that globalisation has led to significant change in China
- To be able to present information using a concept map

Key terms (see glossary)

- capitalism
- communism

Online resources

- **Information sheet**
  8 Background information on China
- **Activity sheet**
  5 Concept mapping globalisation's effect on China
- **PowerPoint**
  6 How is globalisation changing China?
- **Photo set**
  3 Images of China
- **Weblinks**
  12 Geographical Association 'China images'
  13 Concept Mapping
  14 CMAP

Stimulus

Students question the first image on **PowerPoint 6** using the five Ws (what, why, where, when, who?). What might the answers to their questions be?

Development

Introduce the terms capitalism and communism. Have students come across these terms before?

Use **PowerPoint 6** to set the scene for the lesson. This includes a map of China, information about how China has been governed over the centuries (including the Great Leap Forward and the Cultural Revolution), plus some facts and figures about land area, rural/urban populations and the economy (this is also shown on **Information sheet 8**).

Using the information on **PowerPoint 6**, introduce students to the effects of globalisation in China. Discuss how these effects can be categorised as political, economic, social and environmental. Use the examples on **PowerPoint 6** to demonstrate how students can use concept mapping to explore the complex relationships between these factors. Ask them to construct their own concept map to show how globalisation is changing China – either structuring it themselves or using the cards on **Activity sheet 5** as a starting point. They can refer to **Photo set 3**. Some students may wish to go further and consider the difference between economic and cultural globalisation.

Consolidation

Discuss how the lives of Chinese people have become more 'westernised' due to globalisation. What are the benefits of this? Why might some people be unhappy about these changes? Which aspects of globalisation have not been adopted (e.g. freedom of use of the internet)?

Explore how useful the students have found concept mapping as a technique for organising their thoughts and developing their understanding of globalisation in China. Can they think of other issues that could be explored using this technique? Would concept mapping be a useful revision tool?

Extension

Students could explore more images from the 2007 GA China Study Tour (**Weblink 12**) and learn more about concept mapping (**Weblinks 13 and 14**).

Living Geography

Students may consider China to be remote from their own experiences, but should have begun to develop an understanding of how closely our lives are linked to China through the global division of labour (as discussed in previous lessons). Students begin to understand how globalisation has increased the similarities between our lives and the lives of people in China and consider the pros and cons of this.

Exam skills

Concept mapping is a useful skill for students to develop, encouraging them to explore how different issues are linked and to think creatively. This approach can also be a useful way to revise a topic or issue.

Exam practice

With reference to an example you have studied, explain the economic and cultural benefits that globalisation can bring to a country. (6 marks)

(Answers on **Exam practice questions and answers** download.)

Lesson 7

Does the UK need the protection of the EU?

Key questions

- What is the EU and who are its members?
- What is a trading bloc and how does it work?
- How does being an EU member protect the UK's trade?
- How does this affect trade in other parts of the world?

Learning objectives

- To know what the European Union (EU) is and which countries are members
- To understand how the EU operates as a trading bloc to protect the trade of member countries
- To understand that actions taken by the EU affect people and places elsewhere.

Key terms (see glossary)

- European Union (EU)
- protectionism
- quota
- subsidy
- tariff
- trading bloc

Online resources

- **Information sheet**
  9 UK sugar production and EU protection
- **Activity sheet**
  6 EU and global sugar trade
- **PowerPoint**
  7 Does the UK need the protection of the EU?
- **Weblinks**
  15 Globalisation 101
  16 Independent 'Chasing a sugar rush'
  17 EU flags game

Stimulus

Students list as many EU member countries as they can. Reveal the answers on **PowerPoint 7**. Ask students to stand up if they scored 10 or more, then count upwards, with students sitting down as the count goes up. Applaud the students with the highest score.

Development

Using **PowerPoint 7**, introduce students to factors in the EU sugar trade and explain that they are to take part in a role play activity. As the EU representatives who make decisions about the sugar industry in Europe, they must decide how the sugar trade should be organised to provide most benefit to EU countries (**Information sheet 9**). Working in pairs they consider the options on **Activity sheet 6** and complete the consequences grid to show which option they will choose, whether it is free trade or a barrier to trade, and what consequences taking their choice would have. If they can, they should say if their choice involves a tariff, a quota, a subsidy or none of these. You will need to provide definitions for the key terms, such as trading bloc, subsidy, using **PowerPoint 7**.

Once they have completed the

Barges on the River Rhine. Photo: Ollo/iStock

The European Parliament in Brussels. Photo: David Martyn/iStock

grid on **Activity sheet 6**, discuss their decisions for each scenario as a class. This will allow you to check that they have understood that their decisions should benefit EU countries. Students then reflect on the possible impact of their decisions on people and places outside the EU. Questions could include:

- How would you organise the sugar trade in the EU?
- What consequences would this have?
- Is protectionism good for the UK?
- Is it good for the world as a whole?
- Do you think the UK needs the protection of the EU?

Consolidation

Students complete the last part of **Activity sheet 6** individually. They draw their own conclusion about whether protectionism is a good or a bad thing for the UK and for the world.

Extension

Using **Weblinks 15 and 16** students find out more about European sugar policy and to set it in a global context. The EU flag matching game on **Weblink 17** will provide them with a little light entertainment!

Living Geography

The media often claims that young people are not interested in politics, but when I taught this lesson the students found it fascinating! Most of them needed repeated reminding to tailor their decisions to the needs of EU members, as they wanted to 'be fair to everyone'.

Exam skills

This lesson develops areas of geographical vocabulary that students may not encounter every day. It also involves making informed geographical decisions.

Exam practice

Using an example, explain what is meant by the term 'trading bloc'. (4 marks)

(Answers on **Exam practice questions and answers** download.)

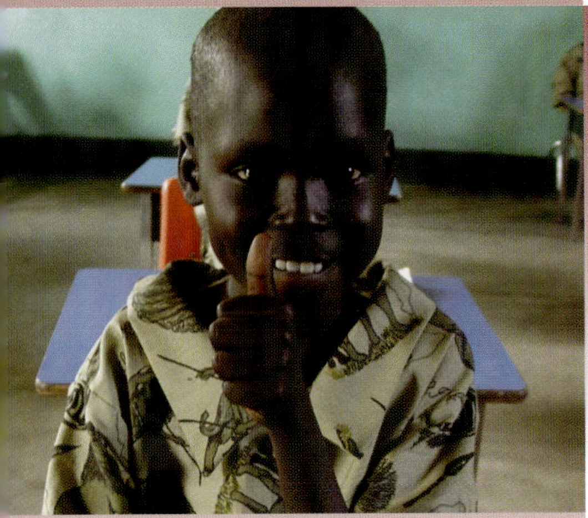

Lesson 8

Are Kenyans winners or losers in the

Key questions

- What is life like for people in Kenya?
- How does globalisation impact upon their daily lives?
- How can globalisation be managed to maximise benefits and minimise problems in places like Kenya?

Learning objectives

- To know where Kenya is and some background information about it
- To understand that different people in Kenya experience globalisation in different ways
- To understand that, with careful management, the benefits of globalisation can be maximised and the problems minimised

Key terms (see glossary)

- continuum
- management
- sustainable

Online resources

- **Information sheets**
  - **10** Kenyan role cards 1-4
  - **11** Kenyan role cards 5-8
  - **12** Kenyan role cards: 9-12
  - **13** Mark scheme
- **Activity sheet**
  - **7** Kenyans: winners or losers in the globalisation game?
- **PowerPoint**
  - **8** Are Kenyans winners or losers in the globalisation game?
- **Photo set**
  - **4** Images of Kenyan people
- **Weblinks**
  - **18** BBC news report 1
  - **19** BBC news report 2
  - **20** Global gang
  - **21** GA 'Kenya images'

Stimulus

Divide students into 12 groups, and give each group a role card from **Information sheet 10**, **11** or **12**. These cards include photos (see also **Photo set 4** and **Weblink 21**) and statements about the lives of 12 Kenyans. Students read the information on the card and arrange themselves into a continuum around the room to show who has the most positive experience of globalisation and who the least. Once they have decided on their place in the continuum, they should each read their card to the class and check if the others agree on their position. They should be aware that there is no 'right' answer and that different people may perceive the same situation in different ways.

Development

Introduce Kenya using the map of Kenya and background information from **PowerPoint 8**. Ask the students to talk about the impact of globalisation in Kenya – use the images of Kenyans on **Information sheets 10–12**.

Introduce the idea of managing globalisation to minimise problems and

The Machahos women's co-operative. Photo: Kevin Cook

lobalisation game?

A new classroom in a school in Marich Pokot.

A supermarket in Eldoret. Photos: Kevin Cook

maximise benefits. Students answer the exam-style question on **Activity sheet 7**, to show how they think the benefits of globalisation in Kenya could be maximised and the problems minimised.

Consolidation

Share examples of answers as a class. Discuss strengths and how weaknesses could be improved. Students peer assess each other's answers using the mark scheme on **Information sheet 13**.

Extension

Students access BBC reports (**Weblinks 18 and 19**) to find out more about Kenya's economy, and play the excellent game on sustainable aid at **Weblink 20**.

Living Geography

In previous lessons, students have looked at globalisation today in relation to China and the UK, whereas this lesson takes them a step further. First they look at how Kenya is currently affected by globalisation and then are asked to consider how globalisation could be managed in the future to benefit the people of Kenya.

Exam skills

As students approach the end of this series of lessons it is useful for them to apply the knowledge, understanding and skills they have developed in an exam-style context.

Exam practice

As the main task for this lesson is an exam-style question, no additional exam practice question is included.

(Answers on **Exam practice questions and answers** download.)

Lesson 9
What will our global future be? Part

Key questions
- What did people imagine our future would be like?
- What is the most likely global future?
- What future would you prefer?

Learning objectives
- To understand the different ways that the future can be envisioned
- To understand that decisions made today will influence the way globalisation affects us in the future

Key terms (see glossary)
- dystopian future
- most likely global future
- preferred global future
- sustainable
- utopian future

Online resources
- **Activity sheet**
  8 Our global futures: preparation
- **PowerPoint**
  9 What will our global future be?
- **Photo set**
  5 Future scenarios
- **Weblinks**
  22 Marxists 'News from Nowhere'
  23 Aldous Huxley 'Brave New World'
  24 George Orwell '1984'
  25 Audacity software

Stimulus
Display the image of baby Megan on **PowerPoint 9** as students enter the room. Tell them Megan was born in 2006. What do they think the world will be like in 2050 when Megan is 44?

Development
Ask students read to read excerpts from what are termed 'utopian' (e.g. **Weblink 22**) and 'dystopian' texts (e.g. **Weblinks 23 and 24**) about how people have imagined the future holds for the world. What do students think the future holds?

Introduce the class to the ideas of most likely and preferred futures on **PowerPoint 9**. Students work in small groups to produce their view of the most likely global future, their preferred future and the steps needed to make their group's preferred future come true. Groups should consider the importance of sustainability when they decide on their preferred future.

Explain that they can present their preferred future as a presentation, a mini-movie, a piece of drama, a sound recording (using the software at **Weblink 25**), a series of images with accompanying notes or as a report/essay. Allow them to choose their own type of presentation as this will enable them to suit their preferred learning style and should result in some really impressive outcomes. Encourage them to take risks and try out new ideas.

Consolidation

Students reflect on what they have achieved during this lesson. Groups use **Activity sheet 8** to record what they need to do in order to be ready to present their findings and preferred future in the next lesson.

Molecular gastronomy. Photo: Thomas Eye Design/ iStock

Living Geography

Over this unit, students will have developed their personal geography in relation to globalisation. The final lessons allow them to demonstrate and celebrate their enhanced personal geography.

Exam skills

Many awarding bodies offer students the opportunity to show their geographical knowledge and understanding in a variety of ways through controlled assessment. For example, the OCR A specification allows students to use 'creative methods to present their work (reflective journals, poems and videos)'. Preparing their preferred future presentations gives students the opportunity to try out different approaches to their work.

Exam practice

What does 'sustainable' mean? Give one example to support your answer. (2 marks)

(Answers on **Exam practice questions and answers** download.)

Lesson 10

What will our global future be? Part

Key questions

- How do our global futures compare?
- What are the strengths of our preferred global futures?
- What are the weaknesses of our preferred global futures?

Learning objectives

- To understand that decisions made today will influence the way globalisation affects us in the future

Key terms (see glossary)

- most likely global future
- preferred global future
- sustainable

Online resources

- **Activity sheet**
  9 Our global futures: presentation and evaluation
- **PowerPoint**
  10 What will our global future be?
- **Weblink**
  26 BBC 'Visions of the future'

Stimulus

Use **PowerPoint 10** to check that groups of students have completed all aspects of the task started in **Lesson 9** and know how they are going to present it to the class.

Development

Each group presents to the class their views of the most likely future, their preferred future and the steps needed to make the preferred future come true. The students assess their peers' presentation using 'stars and wishes' – stars are the strengths of the presentation and wishes are areas for development. They also assess their own presentation using two stars and a wish and record it on **Activity sheet 9**. (Provide a prize and ask the class which group deserves it most.)

If time allows, watch the 'Visions of the future' videos on **Weblink 26**. How are these

Prairie sunrise. Photo: Mills Rymer/iStock

similar to and different from the groups' visions?

Consolidation

Students reflect on what they have learnt through this unit on globalisation. They could give their feedback by writing their views on sticky notes and sticking them to the door as they leave the classroom.

Extension

Students can keep up to date with what is happening in the world in relation to globalisation and interdependence by reading newspapers (in print or online), watching/listening to the news and watching documentaries. Provide some wallspace in your classroom or a page on your department website for them to post links to articles or information on globalisation. Ask them to include a sentence justifying their choice.

Living Geography

The emphasis on 'geographical futures' in this lesson encourages students to apply their understanding of current trends to identify alternative future scenarios and to consider sustainability. It also involves students in thinking about the role they need to play as global citizens to achieve the future they wish for.

Exam skills

In producing their presentations students will need to draw and justify conclusions and present their arguments to an audience. In peer- and self-assessing their presentations they will have the opportunity to discuss the validity and limitations of their conclusions. This also contributes to 'reflective learners' requirements of personal learning and thinking skills.

Exam practice

Choose one of the following issues:

- global climate change
- global food security
- global population growth.

Explain how decisions made today may influence the consequences of this issue in the future. (6 marks)

(Answers on **Exam practice questions and answers** download.)

Quick guide to lesson resources

Lesson 1

PowerPoint
1 What is my place in the globalised world?

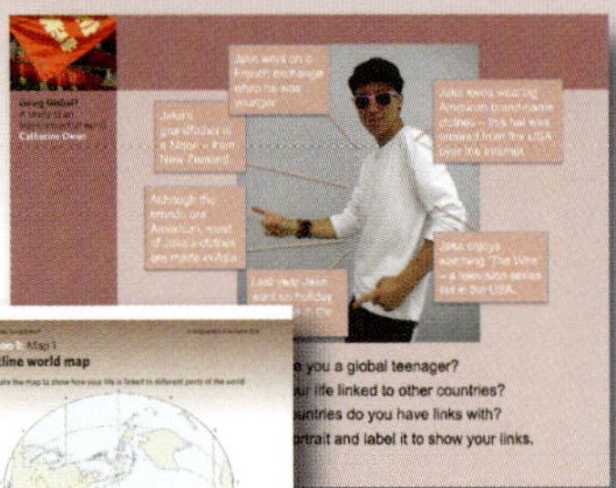

Map
1 Outline world map

Photo set
1 Clothing labels

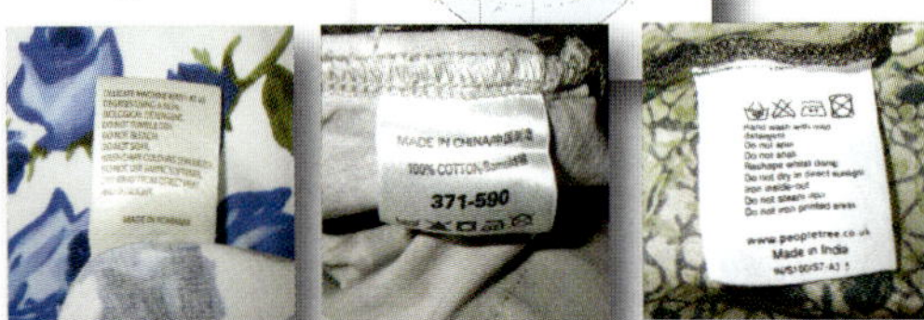

Lesson 2

Information sheet
1 Using Worldmapper

Activity sheet
1 Where does my iPod come from?

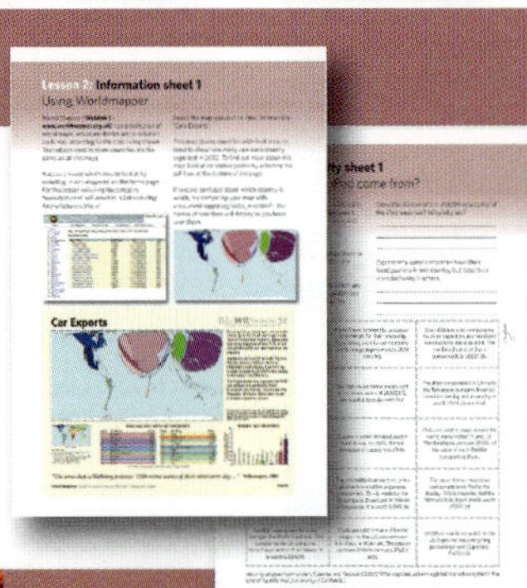

PowerPoint
2 Who made my iPod?

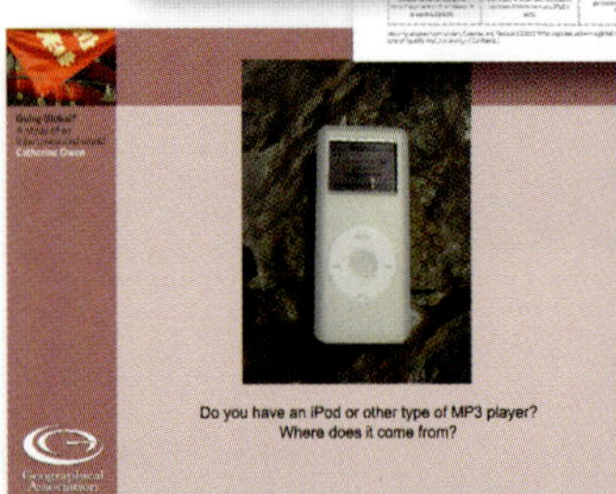

Lesson 3

Information sheets
2 internet use extracts: group A
3 internet use extracts: group B

Activity sheet
2 internet use and globalisation

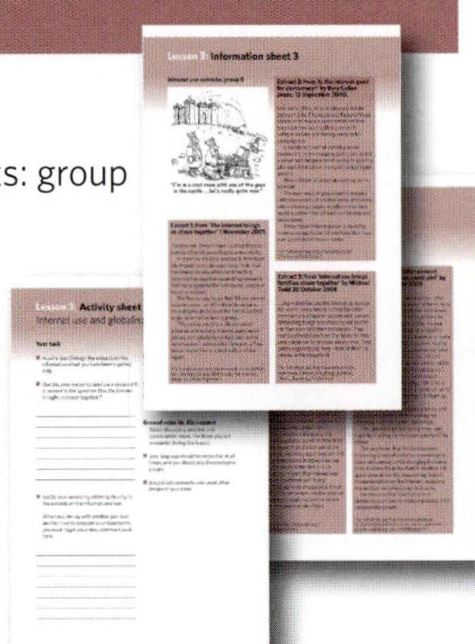

PowerPoint
3 Has the internet brought us closer together?

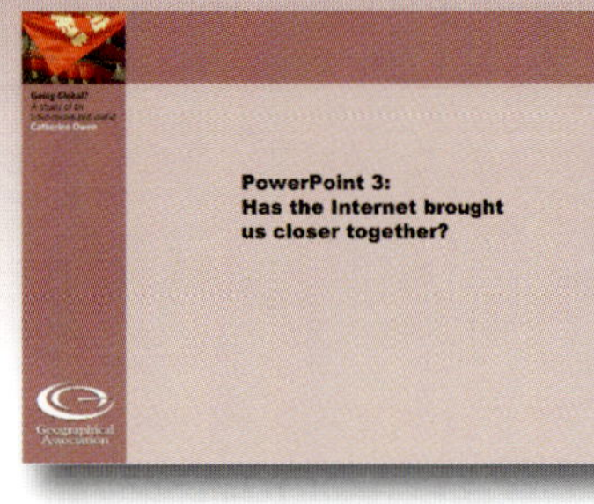

Lesson 4

Information sheet
4 Labels for de Bono's hats

Activity sheet
3 Investigating TNCs using de Bono's hats

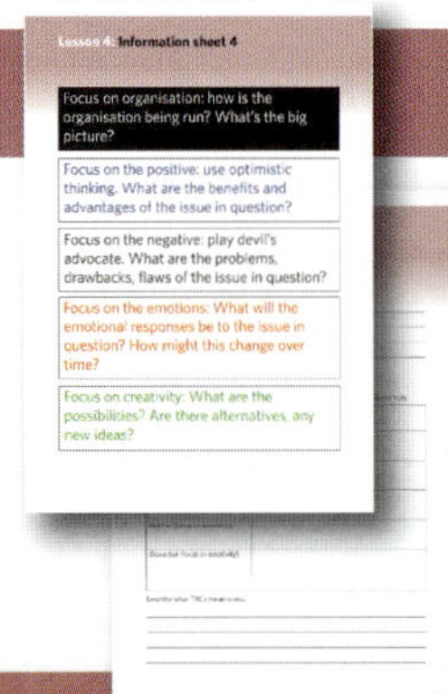

PowerPoint
4 What do TNCs mean to me?

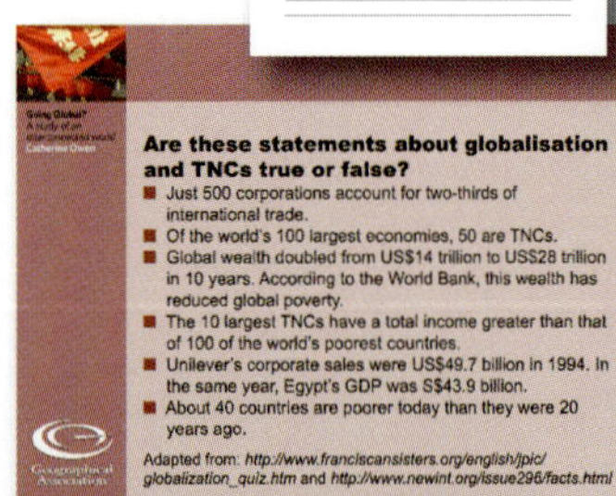

Photo set
2 Kibera, Nairobi, Kenya

Lesson 5

Information sheets
5 Sunbrella simulation: role cards
6 Sunbrella simulation: country labels
7 Sunbrella simulation: price list

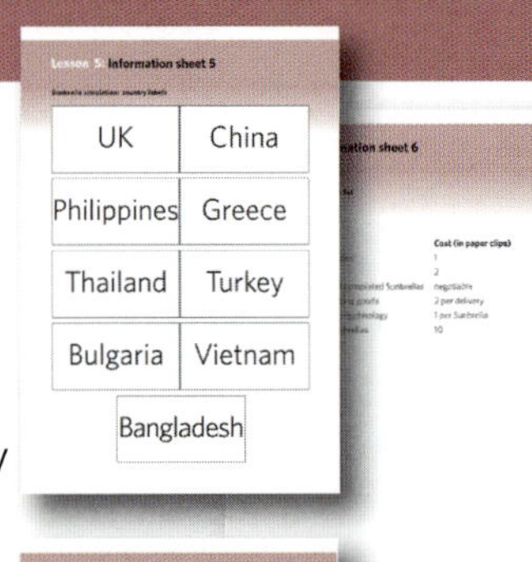

Activity sheet
4 Sunbrella simulation: accounts sheet

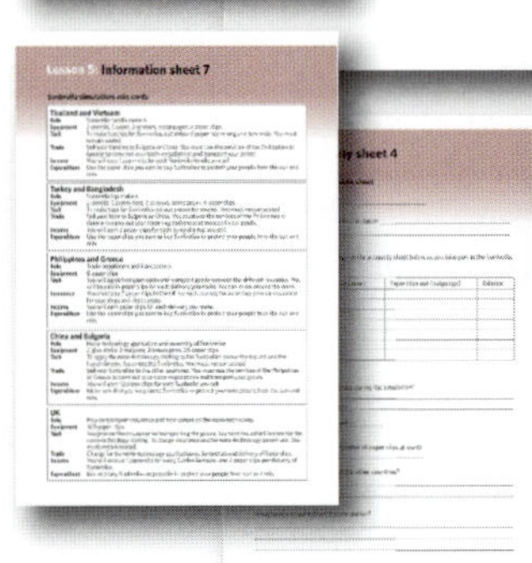

You can access all these resources and more, plus all the weblinks, on the GA website. To download your resources, log on to *www.geography.org.uk/gcsetoolkit*, then click on the button for this book. You will be asked for your password. The unique password for this book is CO25K2.

PowerPoint
5 Global trade – is the balance right?

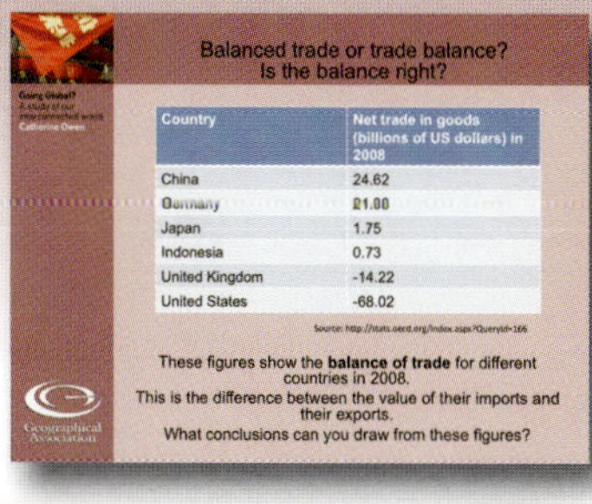

Lesson 6

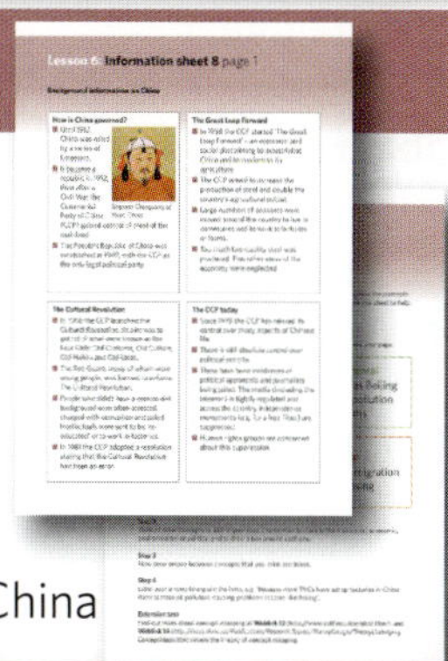

Information sheet
8 Background information on China

Activity sheet
5 Concept mapping globalisation's effect on China

PowerPoint
6 How is globalisation changing China?

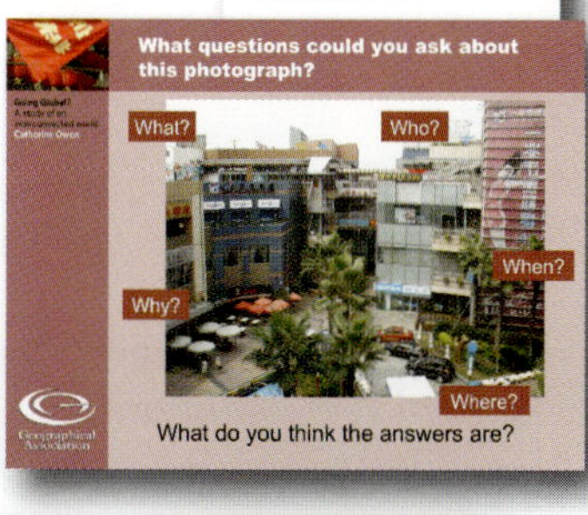

Photo set
3 Images of China

Lesson 7

Information sheet
9 UK sugar production and EU protection

Activity sheet
6 EU and global sugar trade

PowerPoint
7 Does the UK need the protection of the EU?

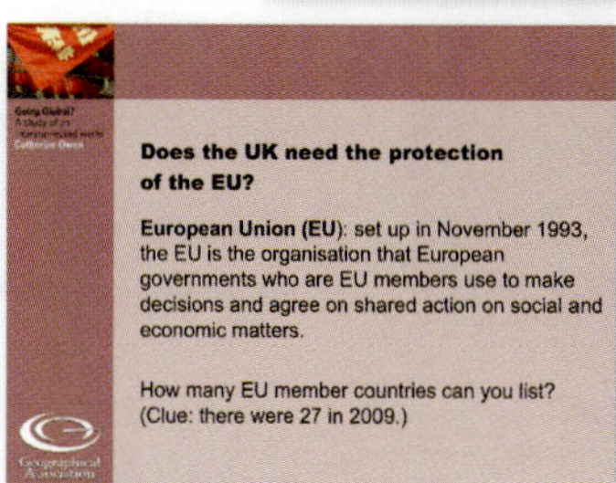

Lesson 8

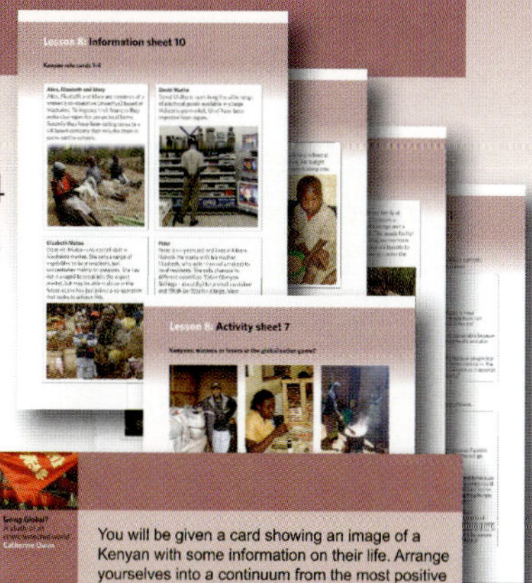

Information sheets
10 Kenyan role cards: 1-4
11 Kenyan role cards: 5-8
12 Kenyan role cards: 9-12
13 Mark scheme

Activity sheet
7 Kenyans: winners or losers in the globalisation game?

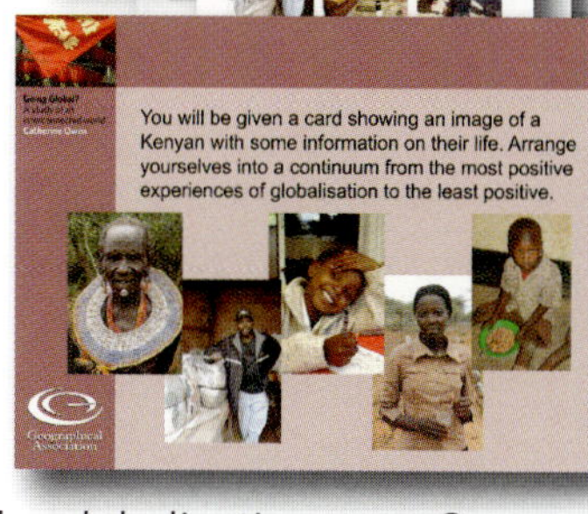

PowerPoint
8 Are Kenyans winners or losers in the globalisation game?

Photo set
4 Images of Kenyan people

Lesson 9

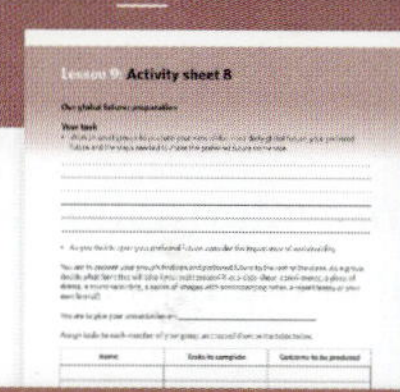

Activity sheet
8 Our global futures: preparation

PowerPoint
9 What will our global future be?

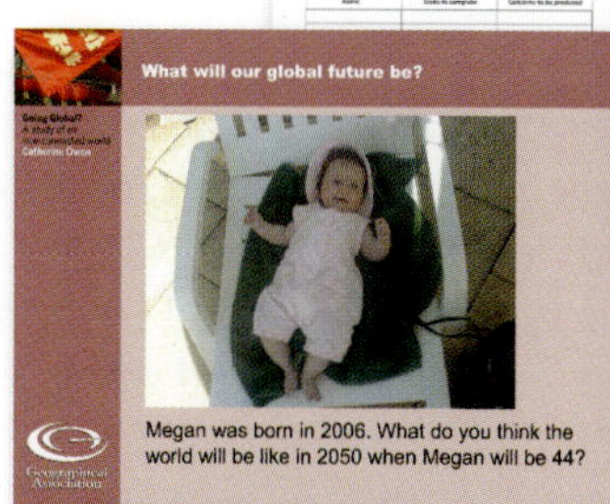

Photo set
5 Future scenarios

Lesson 10

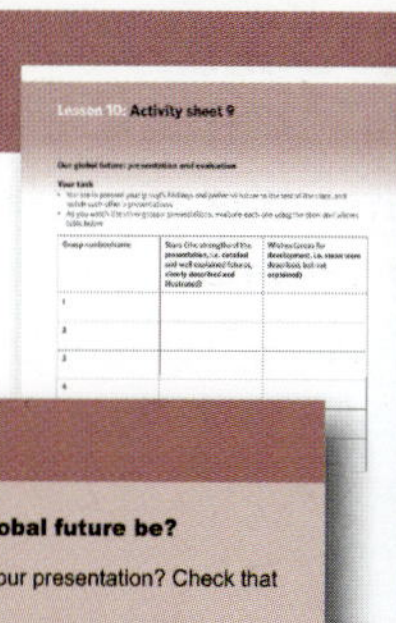

Activity sheet
9 Our global futures: presentation and evaluation

PowerPoint
10 What will our global future be?

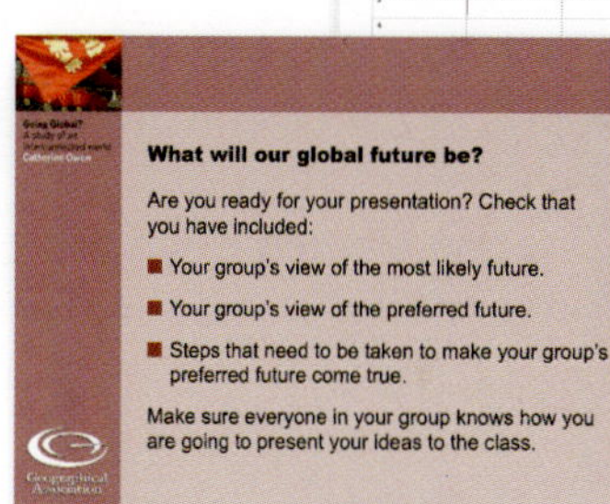

Further teaching ideas

1. What is my place in a globalised world?

Produce a wall display to show which countries your school is linked with and how it is linked with them. Include as many photographs as possible.

Is your school already linked to schools in other parts of the world? Why not ask a group of geography students to come up with ideas for developing these links to support their learning?

Find out about school linking programmes (see e.g. Plan Schools Link and Global Gateway weblinks).

2. What impact does globalisation have on countries at different stages of development?

Arrange a lunchtime/after-school showing of 'The Corporation' (see weblink for more information and ideas for using the film in the classroom).

Would your school be interested in holding a China day? You could run it as an event for gifted and talented students. (See Teaching Expertise and Learning and Teaching Scotland weblinks for ideas about teaching about China.)

Do your students know enough about the EU? Order free diaries with EU information for year 11 students from the Europe Diary website.

Encourage students to run a fundraising lunch and send the profits to the Lunch Bowl project.

3. What will our global future be?

Consider ways to share the best class presentations about global futures with a wider audience, either in assembly or videoed and made available via the school VLE or website.

Watch the 'Visions of the future' videos on the BBC website (**Weblink 26**). How are these similar to and different from your students' visions?

Consider what your students would like to do to make their preferred futures come true. Could you publicise their ideas as a poster campaign? A 'think global and act local' approach could link with sustainability projects in your school (see *www.eco-schools.org.uk*).

CPD ideas

Globalisation as a concept runs through many of the topics taught in geography lessons. During an in-service training session, each teacher in the department could take one topic they already teach and review it to strengthen the globalisation elements and bring the topic up to date. Or you could assess each other's delivery of the lesson with the same outcome in mind.

Sources and weblin

Weblinks used in lesson plans

Weblink 1 - Worldmapper
Equal area world maps on different topics: *www.worldmapper.org.uk*

Weblink 2 - Global gang
Animated game of snakes and ladders including questions about world trade: *www.globalgang.org.uk/funandgames/games/sneakysnakes/snakes.htm*

Weblink 3 - Queens University
Advice on setting ground rules for classroom discussions: *www.queensu.ca/ctl/goodpractice/help/discussion_controversial.html*

Weblink 4 - Eric digests
The importance of holding classroom discussions on controversial issues: *www.ericdigests.org/pre-9218/issues.htm*

Weblink 5 - 'Geography of the internet'
Movie: *www.gatm.org.uk/geographyatthemovies/other.html*

Weblink 6 - 'Shift happens'
A presentation on internet use: *http://shifthappens.wikispaces.com*

Weblink 7 - Instructables
Instructions on making de Bono's hats: *www.instructables.com/id/How-to-Make-a-Paper-Hat*

Weblink 8 - Guardian 'Academic experts'
de Bono's ideas: *www.guardian.co.uk/education/2007/apr/24/highereducationprofile.academicexperts*

Weblink 9 - BBC special report
de Bono's ideas: *http://news.bbc.co.uk/1/hi/special_report/1998/11/98/e-cyclopedia/226700.stm*

Weblink 10 - International Monetary Fund
World trade simulation: *www.imf.org/external/np/exr/center/students/trade/index.htm*

This page can be downloaded from the online resources. It includes hyperlinks to the websites listed.

Weblink 11 - Oxfam 'Milking it'
World trade simulation:
www.oxfam.org.uk/education/resources/milking_it/milkingit/activities/ytf.htm

Weblink 12 - Geographical Association
Images of China with captions:
www.geography.org.uk/events/studytours/china2007 (look also at other pages in this section)

Weblink 13 - Concept Mapping
Information about concept mapping:
www.cotf.edu/ete/pbl2.html

Weblink 14 - CMAP
The theory underlying the use of concept mapping:
http://cmap.ihmc.us/Publications/ResearchPapers/TheoryCmaps/TheoryUnderlyingConceptMaps.htm

Weblink 15 - Globalisation 101
The EU sugar policy and its global context: *www.globalization101.org/index.php?file=news1&id=105*

Weblink 16 - Independent 'Chasing a sugar rush'
Article about the EU sugar policy:
www.independent.co.uk/news/business/analysis-and-features/chasing-a-sugar-rush-global-deficit-drives-price-rises-1770349.html

Weblink 17 - The EU flags game
A flag matching activity:
www.doc.ic.ac.uk/~ih/teaching/game/msc98

Weblink 18 - BBC news report 1
Report on Kenya's economy:
http://news.bbc.co.uk/1/hi/business/3648817.stm

Weblink 19 - BBC news report 2
Report on Kenya's economy:
http://news.bbc.co.uk/1/hi/business/7168060.stm

Weblink 20 - Global Gang
Sustainable aid game:
www.globalgang.org/funandgames/games/gogoatgo/game.html

Weblink 21 - Geographical Association 'China images'
Images of Kenya with captions:
www.geography.org.uk/events/studytours/kenya2006 (look also at other pages in this section)

Weblink 22 - Marxists 'News from Nowhere'
A utopian view of the future:
www.marxists.org/archive/morris/works/1890/nowhere/nowhere.htm

Weblink 23 - Aldous Huxley 'Brave New World'
A dystopian view of the future:
www.huxley.net/bnw

Weblink 24 - George Orwell '1984'
A dystopian view of the future:
www.george-orwell.org/1984

Weblink 25 - Audacity software
Downloadable software for preparing sound recordings:
http://audacity.sourceforge.net

Weblink 26 - BBC 'Visions of the future'
Three future visions for comparison:
www.bbc.co.uk/bbcfour/documentaries/features/visions-future.shtml

Related weblinks

General
BBC 'Talking Point' -
People's comments on the question 'Globalisation: good or bad?':
http://news.bbc.co.uk/1/hi/talking_point/1444930.stm

Economic globalisation
BBC - 'special report into globalisation': *http://news.bbc.co.uk/1/hi/in_depth/business/2007/globalisation/default.stm*

'The Corporation' - a film about economic globalisation:
www.thecorporation.com
Guardian - globalisation:
www.guardian.co.uk/world/globalisation

Globalisation of industry/trade
CDRB - explores global shift of the Bangladesh garment industry:
www.cdrb.org/journal/2004/1/3.pdf

Global issues - criticises WTO for bias towards large corporations:
www.globalissues.org/article/42/the-wto-and-free-trade

World Trade Organisation - role in global trade: *www.wto.org*

EU information
Europe Diary - free diaries with EU information:
www.europadiary.eu/page.asp

China
Learning and Teaching Scotland - ideas for teaching about China:
www.ltscotland.org.uk/mfle/confuciusclassrooms/index.asp

Teaching Expertise - ideas for teaching about China:
www.teachingexpertise.com/articles/lessons-china-3519

Kenya
TIGweb - the concept of globalisation as neo-colonialism:
www.tigweb.org/express/panorama/article.html?ContentID=4363

Yale Global - argues that globalisation (in relation to the film industry) strengthens local cultures:
http://yaleglobal.yale.edu/display.article?id=9655

Glossary

This page can be downloaded from the online resources. It can be used for student reference and revision purposes.

balance of trade - the difference between the money that a country receives from exports and what it spends on imports

capitalism - an economic, political and social system based on private ownership of property, business and industry. It is directed towards making the greatest possible profits for successful organisations and people

communism - the belief in a society without different classes in which the methods of production are owned and controlled by all its members and everyone works as much as they can and receives what they need

continuum - something that changes in character gradually or in small stages without clear dividing points

cultural - relating to the habits, traditions and beliefs of a society

dystopian future* - a vision of a future world in which people live in fear and engage in conflict for idealistic or political reasons or over the resources that are available

economic - relating to trade, industry and money

European Union (EU) - the organisation (since November 1993) through which European member governments make decisions and agree on shared action in social and economic matters

export - to send goods to another country for sale

global division of labour* - a way of organising work, especially making things, so that it is done as a set of separate processes by different (groups of) people

globalisation - when available goods and services, or social and cultural influences, gradually become similar in all parts of the world

import - to buy or bring in products from another country

interdependence - process of being dependent on each other

internet - the network of connected computers around the world which allows people to share information and communicate with each other using, e.g. e-mail

LEDC* - less economically developed country

management - the control and organisation of resources (e.g. people, things or processes)

MEDC* - more economically developed country

media - newspapers, magazines, radio, television and the internet considered as a group

most likely global future* - the scenario that has the greatest likelihood of coming true in the future

multinational - a company that produces and sells goods in many different countries

newly industrialised countries (NICs)* - countries that were once LEDCs, but have now become industrialised (e.g. Taiwan, Malaysia)

preferred global future* - the scenario that someone would like to come true in the future

primary sector* - industries that involve harvesting natural resources, such as farming, mining and fishing

protectionism - the actions of a government to help its country's trade or industry by taxing goods bought from other countries

quaternary sector* - industries that include intellectual services: information generation, information sharing, consultation, education and research and development

quota* - a limit set on production to avoid surpluses, e.g. milk quotas limit the amount of milk farmers are allowed to produce.

secondary sector* - industries that involve manufacturing something or assembling components

subsidy* - money given, usually by government, to encourage the development of a product or project which might not otherwise be economically viable

sustainable - causing little or no damage to the environment and therefore able to continue for some time

tariff - a charge or list of charges either for services or on goods entering a country

tax - (an amount of) money paid to the government, which is based on your income or of the cost of goods or services you have bought

tertiary sector* - industries that provide a service, such as banks, shops and schools

trading bloc* - a type of inter-governmental agreement where regional barriers to trade are reduced or eliminated among the participating states

transnational corporations (TNCs)* - large business organisations involving management and production of goods in several nations

utopian future* - idealistic (imagined perfect) vision of a future world

All definitions taken from *http://dictionary.cambridge.org* unless marked with *.